Hepatoprotective Models and Natural Hepatoprotective Agent

Liver is the largest and most vital organ of the body. It is also called as metabolic 'engine room' of the body. Most of the drugs, food and water constituents get metabolized and detoxified in the liver and as it are often exposed maladies and it will result in number of clinical syndromes. Variety of liver disease such hepatitis, jaundice, cirrhosis, liver cancer etc caused by the many of chemicals, foods, drugs and infections parasitic, bacterial, fungal and viral because of variations in liver dysfunction and difficulties encountered in reaching to a proper diagnosis , a physician is unable to provide proper treatment. Liver involved in several vital functions such as metabolism, secretion, and storage. As well as detoxification of several drugs and xenobiotics occurs in the liver itself.

Among the disorders liver diseases is the most serious. These can be classified as acute or chronic hepatitis (inflammatory liver diseases), hepatosis (non-inflammatory liver diseases) and cirrhosis (degenerative disorder cause fibrosis of the liver. Toxic chemicals such as (certain antibiotics, chemotherapeutics, peroxidised oil, aflatoxin, carbon tetrachloride, acetaminophen, chlorinated hydrocarbons, etc) are responsible for liver diseases. Others are excess consumption of alcohol and auto-immune disorders. Liver microsomal metabolism of ethanol may result in hepatitis and cirrhosis due to enhanced lipid peroxidation. It has been found that viruses are responsible for 90% acute hepatitis. Conventional medicines are now pursuing the use of natural products such as herbs to provide the support that liver needs on a daily basis.

Hepatoprotective Model for Animal Study: Both *in vivo* and *in vitro* models are employed for assessing the hepatoprotective activity. Both these models measure the ability of the test drug to prevent or cure liver toxicity which may be induced by various hepatotoxins in experimental animals such as rats, mice etc.

In vitro **models:** For studying the antihepatotoxic activity of drugs, fresh hepatocyte preparations and primary cultured hepatocytes are cultured and are treated with hepatotoxin. The effect of the test drug on the hepatotoxin treated cultured hepatocytes is evaluated. The transaminases activities released into the medium are determined. Liver damage is indicated by an augmented activity of marker transaminases in the medium. Parameters determined are hepatocytes multiplication, morphology, macromolecular synthesis and oxygen consumption.

In vivo models: Liver damage in experimental animals is induced by administration of a toxic dose or repeated doses of a known hepatotoxin. The test substance is administered along with, prior to and/or after the toxin treatment. Liver damage and recovery from damage are assessed by quantifying serum marker enzymes, bilirubin, bile flow, histopathological changes and biochemical changes in liver. Liver damage is indicated by an augmented activity of liver marker enzymes such as glutamate pyruvate transaminase (GPT), glutamate oxaloacetate transaminase (GOT) and alkaline phosphatase in the serum.

Hepatotoxic agents: Various chemical agents normally used to induce hepatotoxicty in experimental animals for the evaluation of hepatoprotective agents are Acryl amide, Adriamycin, Alcohol, Antitubercular drugs, Cadmium, Carbon tetrachloride, Erythromycin, Galactosamine, Lead, Microcystin, Paracetamol, Tamoxifen, Thioacetamide, etc.

Acryl amide: Acryl amide (AA) is a water-soluble vinyl monomer and is carcinogenic to humans. In the human body, AA is oxidized to the epoxide glycidamide (2,3-epoxypropionamide) through an enzymatic reaction involving cytochrome P4502E1. AA undergoes biotransformation by conjugation with glutathione and is probably being the major route of detoxification. Daily dose of 6 mg/kg, ip for 15 days is used for the production of hepatotoxicity in female Sprague-Dawley rats.

Adriamycin: Adriamycin (Doxorubicin) is a potent cytotoxic agent which has been shown to undergo redox cycling between semiquinone and quinone radicals during its oxidative metabolism. A single dose of 10 mg/kg body weight of adriamycin is given to rats for inducing hepatotoxicity.

Alcohol: Alcohol consumption causes fatty infiltration, hepatitis and cirrhosis of the liver. Fat infiltration is a reversible phenomenon which occurs when alcohol replaces fatty acids in the mitochondria. Hepatitis and cirrhosis occurs because of enhanced lipid peroxidative reaction during the microsomal metabolism of ethanol. These effects of ethanol are a result of the enhanced generation of oxy free radicals during its oxidation in liver. The peroxidation of membrane lipids leads to loss of membrane structure and integrity. These results in elevated levels of glutamyl transpeptidase, a membrane bound enzyme in serum. Continuous administration of ethanol (7.9 g/kg body weight/d) for a period of 6 weeks causes liver damage in rats.

Antitubercular drugs: Isoniazid is metabolized to monoacetyl hydrazine, which is further metabolized to a toxic product by cytochrome P450 leading to hepatotoxicity. Patients on concurrent rifampicin therapy have an increased risk of hepatitis.

Cadmium: Cadmium exposure causes testicular atrophy, renal dysfunction, hepatic damage, hypertension, central nervous system injury and anemia. Cadmium induces oxidative damage in different tissues by enhancing peroxidation of membrane lipids in tissues and altering the antioxidant systems of the cells. Cadmium is given orally (3 mg/kg body weight/d) as cadmium chloride ($CdCl_2$) for 3 weeks to induce hepatotoxicity in rats.

Carbon tetrachloride: Carbon tetrachloride is metabolized by cytochrome P-450 in endoplasmic reticulum and mitochondria with the formation of CCl_3O^-, a reactive oxidative free radical, which initiates lipid peroxidation. Dose of CCl_4: 1 mL/kg body weight, i.p., 1:1 v/v mixture of CCl_4 and olive oil induces hepatotoxicity.

Erythromycin: Erythromycin estolate is a potent macrolide antibiotic which generates free radicals and induces liver toxicity. Erythromycin when given as erythromycin stearate (100 mg/kg body weight for 14 d) or erythromycin esolate (800 mg/kg/d for 15 d) to albino rats produces hepatotoxicity.

Galactosamine: Galactosamine causes diffuse type of liver injury. Galactosamine decreases the bile flow and its content i.e. bile salts, cholic acid and deoxycholic acid. Galactosamine reduces the number of viable hepatocytes as well as rate of oxygen consumption. Hepatic injury is induced by intraperitoneal single dose injection of D-galactosamine (800 mg/kg).

Lead: Lead-induced hepatic damage is mostly rooted in lipid peroxidation (LPO) and disturbance of the prooxidant-antioxidant balance by generation of reactive oxygen species (ROS). Hepatotoxicity can be induced by using lead acetate (550 ppm for 21 d in drinking water) or lead nitrate (5 mg/kg body weight daily for 30 d).

Microcystin: Microcystis aeruginosa, is a potent hepatotoxin. Mice receiving sublethal doses of microcystin (20 μg/kg) for 28 weeks developed neoplastic liver nodules.

Paracetamol: Paracetamol is a widely used analgesic and antipyretic drug which produces acute liver damage in high doses. Paracetamol administration causes necrosis of the centrilobular hepatocytes characterized by nuclear pyknosis and eosinophilic cytoplasm followed by large excessive hepatic lesion. Dose of Paracetamol that causes hepatotoxicity is 2 g/kg.

Tamoxifen: Tamoxifen citrate (TAM) is a non-steroidal antiestrogen drug used in the treatment and prevention of hormone dependent breast cancer. At high doses, it causes liver carcinogenicity in rats, due to oxygen radical overproduction and lipid per oxidation via formation of lipid peroxy radicals. An ip dose of 45 mg/kg/d of tamoxifen citrate in 0.1 mL dimethylsulfoxide and normal saline for 6 d induce hepatotoxicity in rats.

Thioacetamide: Thioacetamide is reported to interfere with the movement of RNA from the nucleus to cytoplasm which causes membrane injury. A metabolite of thioacetamide is responsible for hepatic injury. Thioacetamide reduces the number of viable hepatocytes and rate of oxygen consumption as well. It also decreases the volume of bile and its content i.e. bile salts, cholic acid and deoxycholic acid. I.P. dose of thioacetamide that causes hepatotoxicity is 200 mg/kg, thrice weekly for 8 weeks.

Clinical research in this century has confirmed the efficacy of several plants used in the treatment of liver diseases. Basic scientific research has uncovered the mechanism by which some plants afford their therapeutic effect. Physicians and patient need effective therapeutic agent with minimum incidence of toxic effects.

1. *Curcuma longa* **(Turmeric):** *Curcuma longa* is a member of the ginger belongs to the family Zingiberaceae. Raw turmeric contains 0.3-5.4 percent curcumin (diferuloylmethane). Turmeric also contains 4-14 percent volatile oils, including tumerone, atlantone, and zingiberone. Turmeric also contains sugars (28 percent glucose, 12 percent arabinose), proteins, and resins. Traditional applications of turmeric which is derived from dried, ground rhizome include the treatment of gastrointestinal colic, flatulence, hemorrhage, hematuria, menstrual difficulties, and jaundice. The anti-inflammatory and hepatoprotective characteristics of turmeric and its constituents have been widely researched. The hepatoprotective effects of turmeric are due to its potent antioxidant effects. Both the volatile oil and curcumin exhibit powerful anti-inflammatory effects. Curcumin also has choleretic

effects on the liver. Based on clinical experience, a typical recommended dose of curcumin is 400-600 mg three times per day.

CURCUMIN

2. *Camellia sinensis* (**Green Tea**): Green, black, and oolong teas derived from of the leaves of *Camellia sinensis* belongs to the family Theaceae, contains a wide range of bioactive constituents, most of which are contained in two groups, alkaloids and polyphenols. Examples of alkaloids found in tea include caffeine, the bromine, and theophylline. The polyphenols contained in teas are classified as Catechins. Green tea contains six primary catechin compounds: (+)-catechin, gallocatechin, epicatechin, epigallocatechin, epicatechin gallate, and epigallocatechin gallate. Epigallocatechin gallate (also known as EGCG) is considered to be the most active component. Historical uses of tea are as a stimulant, an astringent for clearing phlegm, and as a digestive aid. Catechins are powerful antioxidants, which are responsible for green tea's hepatoprotective activity. Pure (+)-catechin (also known as (+)-cyanidanol-3 – trade name Catergen) has been used to treat hepatitis since 1976. Green Tea also shows detoxification activity and anti-cancer properties. Single doses of decaffeinated green tea solids up to 4.5 g/day (equal to 45 cups of tea) have been well tolerated by humans.

CAFFEINE   THEOPHYLLINE   CATECHINS

3. *Glycyrrhiza glabra* (**liquorice**): *Glycyrrhiza glabra* belongs to the family Fabaceae. The primary active constituent of Glycyrrhiza, as it relates to hepatic disorders, is the triterpene

glycoside glycyrrhizin (also known as glycyrrhizic acid or glycyrrhetinic acid) and is derived from the roots (6-14 percent). Other constituents of Glycyrrhiza include flavonoids (liquiritin and isoliquiritin), isoflavonoids (isoflavonol, kumatakenin, licoricone, and glabrol), chalcones, coumarins (umbelliferone, herniarin), triterpenoids, and phytosterols. Traditional uses include the treatment of peptic ulcers, asthma, pharyngitis, malaria, abdominal pain, and infections. The traditional medicinal properties of Glycyrrhiza include demulcent, expectorant, antitussive, and mild laxative activity. Liquorice is used to flavor a wide variety of candies, gum, tobacco products and drinks. The surfactant property of the steroidal saponins may also facilitate absorption of poorly-absorbed compounds, such as carotenes and anthraquinone glycosides. Glycyrrhiza has been shown to have a direct hepatoprotective effect. Glycyrrhiza enhances the detoxification of medications and toxins. Glycyrrhiza exerts antiviral activity in vitro toward a number of viruses, including hepatitis A, varicellazoster , HIV, herpes simplex type 1, Newcastle disease, and vesicular stomatitis viruses. Glycyrrhiza is well tolerated by most patients at normal doses (1-4 g/d crude herb).

GLYCYRRHIZIC ACID

GLYCYRRHETINIC ACID

4. **Terminalia chebula (Indian Gall Fruit):** *Terminalia chebula* belongs to the family Combretaceae. Part used is stem bark and fruit. Researchers have isolated a number of glycosides from Haritaki, including the triterpenes arjunglucoside I, arjungenin, and the chebulosides I and II. Other constituents include a coumarin conjugated with gallic acids called chebulin, as well as other phenolic compounds including ellagic acid, 2,4-chebulyl-β-D-glucopyranose, chebulinic acid, gallic acid, ethyl gallate, punicalagin, terflavin A, terchebin, luteolin, and tannic acid. Chebulic acid is a phenolic acid compound isolated from the ripe fruits.Luteic acid can be isolated from the bark. Traditionally used in chronic diarrhoea and dysentery, flatulence, vomiting, colic and enlarged spleen and liver. Indian gall

fruit has a hypocholesterolemic effect on cholesterol-induced hypercholesterolemia. Dose is 200 and 400 mg/kg, p.o.

COUMARIN

ELLAGIC ACID

5. ***Cichorium intybus* (Chicory Seed):** *Cichorium intybus* belongs to the family Asteraceae. Its leaves are a prime source of two sesquiterpene lactones Lactucin and Lactucopicrin. Other ingredients are Aesculetin, Aesculin, Cichoriin, Umbelliferone, Scopoletin and 6.7-Dihydrocoumarin and further sesquiterpene lactones and their glycosides. Traditionally used for hepatic conditions and liver rejuvenation and has shown protective effects in mice with high levels of liver damaging enzymes. Dose is 50, 100 and 200 mg/kg, i.p.

LACTUCIN

LACTUCOPICRIN

6. ***Piper longum* (Long Pepper Fruit):** *Piper longum* belongs to the family Piperaceae. As an antihypertensive and sedative, Indian Long Pepper is beneficial in treating insomnia. The herb is useful in the treatment of respiratory disorders like the common cough, bronchitis and

asthma. It is also a potent digestive, which helps in treating gastrointestinal disorders like indigestion. Piperine is the major alkaloid found in Indian Long Pepper (fruit). Piperine is antipyretic, hypotensive and a central nervous system stimulant. Piperine has been shown to exert a significant protection against liver toxicity induced by tert-butyl hydroperoxide and carbon tetrachloride by reducing both in vitro and in vivo lipid peroxidation by decreasing the reduction of GSH.

PIPERINE

7. *Caesalpinia bonducella* (**Fever nut**): The seeds, leaves and bark of medicinal plant *Caesalpinia bonducella* belong to the family Caeselpiniaceae are used in fever, asthma, colic. It contains a bitter substance bonducin, phytosterinin, fatty acids, caesalpins (α, β, γ, δ and ψ), new diterpene caesalpins, a new homoisoflavone-bonducelline and citrulline. The seeds of *Caesalpinia bonducella* contain a bitter principle bonducin-2%, fatty oil-25%, proteins-20% and starch-35.5%. Four cassane furanoditerpenes, designated bonducellpins A, B, C & D were isolated; two new cassane diterpenes, named caesaldekarins F & G have been isolated & identified; α-amyrin, β-amyrin, lupeol and lupeol acetate have been identified and isolated. The whole plant is used as emmenagogue, febrifuge, astringent, anthelmintic, digestive, stomachic, liver tonic, depurative, expectorant, antipyretic, aphrodisiac, thermogenic, splenomegaly, hepatomegaly, amenorrhoea, dysmenorrhoea, pharyngodynia & tonic. The seeds are bitter, acrid, anodyne, anti-inflammatory, antifertility, cough, asthma, leucoderma, leprosy, skin diseases, dyspepsia, dysentery, colic, haemorrhoids, hepatopathy, diabetes and intermittent fevers. The leaves are anthelmintic, febrifuge and are useful in elephantiasis, intestinal worms, splenomegaly and fevers .The young leaves are used in hepatic disorders. The fatty oils obtained from the nucleus of the seeds are useful in convulsions and paralysis. *Caesalpinia bonducella* possesses antipyretic and analgesic, hypoglycaemic, antihyperglycaemic and hypolipidemic, and antibacterial activities.

8. *Emblica officinalis* (**Amla**): The fruits of *Emblica officinalis* belonging to the family Combretaceae are reputed to contain high amounts of ascorbic acid (vitamin C), 445 mg/100g, the specific contents are disputed, and the overall antioxidant strength of amla may derive instead from its high density of ellagitannins such as emblicanin A (37%), emblicanin B (33%), punigluconin (12%) and pedunculagin (14%). It also contains punicafolin and phyllanemblininA, phyllanemblin other

    polyphenols: flavonoids, kaempferol, ellagic acid and gallic acid. Indian gooseberry has undergone preliminary research, demonstrating *invitro* antiviral and antimicrobial properties. There is preliminary evidence *in vitro* that its extracts induce apoptosis and modify gene expression in osteoclasts involved in rheumatoid arthritis and osteoporosis. It may prove to have potential activity against some cancers. One recent animal study found treatment with *E. officinalis* reduced severity of acute pancreatitis (induced by L-arginine in rats). It also promoted the spontaneous repair and regeneration process of the pancreas occurring after an acute attack. Experimental preparations of leaves bark or fruit have shown potential efficacy against laboratory models of disease, such as for inflammation, cancer, age-related renal disease, and diabetes. A human pilot study demonstrated a reduction of blood cholesterol levels in both normal and hypercholesterolemic men with treatment. Another recent study with alloxan-induced diabetic rats given an aqueous amla fruit extract has shown significant decrease of the blood glucose, as well as triglyceridemic levels and an improvement of the liver function caused by a normalization of the liver-specific enzyme alanine transaminase activity.

ASCORBIC ACID

9. *Boerhaavia diffusa* (**Spreading Hog Weed**): Boeravinones G and H are two rotenoids isolated from whole plant of *B. diffusa* belonging to the family Nyctaginaceae. *Boerhavia diffusa* is believed to improve and protect eyesight. *B. diffusa* has diuretic

properties and is used by diabetics to lower blood sugar. *Boerhavia diffusa* has shown antibacterial activity, mainly against Gram-negative bacteria. Extracts of *B. diffusa* leaves have shown antioxidant and hepatoprotective properties in pharmacological models. Punarnavine (an alkaloid isolated from *B. diffusa*) has shown some *in vitro* anticancer, antiestrogenic, immunomodulatory, and antiamoebic activity (particularly against *Entamoeba histolytica*). *Boerhavia diffusa* is a source of antioxidants, and may be effective against arsenic trioxide (an effective drug used against acute promyelocytic leukemia) induced cardio toxicity. B diffusa also possess cardioprotective properties.

BOERAVINONES

10. *Terminalia arjuna* (**Arjuna Myrobalan Bark**): *Terminalia arjuna* belongs to the family Combretaceae. The bark of Arjuna tree is used for medicinal purposes and it has been found to contain minerals such as calcium, magnesium, aluminium, and tannins, flavonoids, saponin glycosides and phytosterols. The bark also contains crystalline compounds such as arjunine, arjunetin, essential oils and reducing sugars.

Arjuna bark has anti-inflammatory properties which act as COX inhibitors and non-steroidal anti-inflammatory agents, and displayed both analgesic and anti-inflammatory properties. Studies have shown that Arjuna tree is effective in bringing down LDL cholesterol levels. Arjuna bark has been traditionally prescribed for heart problems. Recent studies have shown that Arjuna was very effective in controlling refractory chronic congestive heart failure. The research concluded Arjuna to be a potent diabetes reducing agent.

The antioxidants present in Arjuna bark acted as nullifying agents against fluoride damage caused to the liver. Arjuna bark extracts were so effective that fluoride levels had come down to almost normal after just 10 days. The anti-oxidants play in Arjuna bark play a major role in scavenging free radicals and minimizing their damage. According to Ayurveda, Arjuna bark can be very effective in the treatment of asthma. Fine powder of the dried bark must be taken with kheer or rice and milk pudding. Arjuna bark powder can also be effective in reducing both diarrhoea and dysentery. According to Ayurveda, Arjuna bark is effective in restoring strength to the bones which have been fractured. Powdered dry bark of Arjuna can be taken along with honey for this.

11. ***Silybum marianum* (Milk Thistle):** *Silybum marianum* belongs to the family Asteraceae. Traditional milk thistle extract is made from the seeds, which contain approximately 4–6% silymarin. The extract consists of about 65–80% silymarin (a flavonolignan complex) and 20–35% fatty acids, including linoleic acid. Silymarin is a complex mixture of polyphenolic molecules, including seven closely related flavonolignans (silybin A, silybin B, isosilybin A, isosilybin B, silychristin, isosilychristin, silydianin) and one flavonoid (taxifolin). Silibinin, a semipurified fraction of silymarin, is primarily a mixture of 2 diastereoisomers, silybin A and silybin B, in a roughly 1: 1 ratio. In clinical trials silymarin has typically been administered in amounts ranging from 420–480 mg per day in two to three divided doses. However higher doses have been studied, such as 600 mg daily in the treatment of type II diabetes and 600 or 1200 mg daily in patients chronically infected with hepatitis C virus. An optimal dosage for milk thistle preparations has not been established. Milk thistle, along with dandelion and other extracts are often referred to as hangover cures as the bitter tincture helps organs rid toxins after heavy drinking.

SILYMARIN

12. ***Picrorhiza kurroa* (Kutki):** *Picrorhiza kurroa* (of the family Scrophulariaceae), also known as Kutki or Katuki, is a perennial herb used in Ayurveda. It is traditionally used for liver disorders, but has also been implicated in the treatment of upper respiratory tract, fevers, dyspepsia, chronic diarrhea, and scorpion stings. It consists of the bitter principle known as 'Kutkin', which is a mixture of picroside I and picroside II (kutkoside). These are irioid glycoside structures present at 1.611% and 0.613% of the roots dry weight, respectively. It also contains α-methoxy substituted catechol Apocynin, structurally similar to vanillic and

ferulic acids, Androsin, Cucurbitacin glycosides based on cucurbitacin B and dihydrocucurbitacinB.

KUTKIN(PICROSIDE2)

KUTKIN(PICROSIDE1)

KUTKOSIDE

13. ***Andrographis paniculata* (Kalmegh):** King of Bitters botanically known as *Andrographis paniculata* belongs to a family Acanthaceae. It is an ancient Indian medicinal herb, which has been used for centuries in Asia for its effects on various bodily functions and ailments, ranging from degenerative diseases to the common cold. It is known as Kalmegh and is used

as a bitter ingredient in the Indian indigenous system of medicine. The leaves contain andrographolide, most active component of *Andrographis paniculata* is very bitter in taste. One the most common therapeutic potential of *Andrographis paniculata* is its liver protective property, which is well established experimentally. Alcoholic extract of the leaves of *Andrographis paniculata* was found to be effective in prevention of liver damage. In another study administration of *Andrographis paniculata* exhibited liver protective effects by enhancing activity of antioxidant enzymes like superoxide dismutase, catalase, glutathione peroxidase, glutathione reductase along with the level of glutathione and decreasing the activity of lipid peroxidase which leads to generation of free radicals damaging the liver cells. Thus by means of its synergistic effects *Andrographis paniculata* exerts its well-known hepatoprotective action.

ANDROGRAPHOLIDE

14. ***Foeniculum vulgare* (Fennel):** Fennel (*Foeniculum vulgare* Mill., family Umbelliferae) is an annual, biennial or perennial aromatic herb, depending on the variety, which has been known since antiquity in Europe and Asia Minor. The leaves, stalks and seeds (fruits) of the plant are edible. *Foeniculum vulgare* is an aromatic herb whose fruits are oblong, ellipsoid or cylindrical, straight or slightly curved and greenish or yellowish brown in colour. Volatile components of fennel seed extracts by chromatographic analysis include transanethole, fenchone, methylchavicol, limonene, α- pinene, camphene, β-pinene, β-myrcene, α-phellandrene, 3-carene, camphor, and cisanethole. Hepatoprotective activity of *Foeniculum vulgare* (fennel) essential oil was studied using a carbon tetrachloride-induced liver fibrosis model in rats. The hepatotoxicity produced by chronic carbon tetrachloride administration was found to be inhibited by *Foeniculum vulgare* essential oil with evidence of decreased

levels of serum aspartate aminotransferase, alanine aminotransferase, alkaline phosphatase and Billirubin.

TRANS-ANETHOLE

FENCHONE

BETA-MYRCENE

15. **Swertia chirata (Bitter Stick):** *Swertia chirata* commonly known as clearing nut tree, bitter stick, Indian chirette, dowa I pechish, Indian gentian is consists of the entire herb of *Swertia chirata*, belonging to family Gentianceae. *Swertia chirata* contains bitter yellow acid known as ophelic acid, two bitter glycosides chiratin (not a pure substance) and amarogentin (phenol carbonic acid ester of sweroside, a substance related to gentiopicrin), two alkaloids gentianine and gentiocrucine and also contains yellow crystalline substance used in dyeing. Apart from hepatoprotective action, the drug also shows digestive, hepatic (conditions pertaining to the liver), tonic, astringent and appetizer properties and used in cough, dropsy and skin diseases.

GENTIANINE

GENTIOCRUCINE

16. **Azadirachta indica (Neem):** Dried leaf of *Azadirachta indica* family Meliaceae consists of triterpenoids, sterols, bitter principles nimbin and nimbiol. Neem bark is cool, bitter, astringent, acrid and refrigerant. It is useful in tiredness, cough, fever, loss of appetite, worm infestation. It heals wounds and vitiated conditions of kapha, vomiting, skin diseases,

excessive thirst, and diabetes. Neem leaves are reported to be beneficial for eye disorders and insect poisons. It treats Vatik disorder. It is anti-leprotic. Its fruits are bitter, purgative, anti-hemorrhoids and anthelmintic. It is claimed that neem provides an answer to many incurable diseases. Traditionally neem products have been used against a wide variety of diseases which include heat-rash, boils, wounds, jaundice, leprosy, skin disorders, stomach ulcers, chicken pox, etc.

NIMBIN

NIMBOL

17. *Eclipta alba* (**Bhringaraja**): The whole plant of *Eclipta alba* Hassk family Asteraceae consists of Alkaloids, Ecliptine and Nicotine. It is used as a deobstruent, antihepatotoxic, anticatarrhal, febrifuge. Used in hepatitis, spleen enlargements, chronic skin diseases. Leaf— promotes hair growth. Its extract in oil is applied to scalp before bed time in insomnia. The herb is also used as an ingredient in shampoos.

.NICOTINE

18. ***Cynodon dactylon* (Doob):** Durva consists of dried fibrous roots of Cynodon dactylon Linn family Poaceae. Its major constituents are Phenolic phytotoxins and flavonoids. In ethnomedicinal practices the juice of the plant *Cynodon dactylon* Pers. is used as astringent and is applied to fresh cuts and wounds. It is used internally in the treatment of chronic diarrhoea and dysentery. It is also useful in the treatment of Catarrhal ophthalmia. The leaves of *Cynodon dactylon* Pers. are also used in the treatment of hysteria, epilepsy and insanity. The plant *Cynodon dactylon* Pers. is also a folk remedy for anasarca, calculus, cancer, carbuncles, convulsions, cough, cramps, cystitis, diarrhoea, dysentery, headache, haemorrhage, hypertension, kidneys, laxative, measle, rubella, sores, stones, tumours, uro-genital disorders, warts and wounds.

19. ***Lagenaria siceraria, L. leucantha, L. vulgaris* (Lauki):** It consists of fresh fruit (devoid of stalk) of Lagenaria siceraria Syn. L. leucantha Rusby L. vulgaris Ser family Cucurbitaceae. Analysis of edible portion of the fruit contains protein, fat (ether extract), carbohydrates, mineral matter like calcium and phosphorus. Glucose and fructose have been detected. The amino acid composition of the fruit is leucines, phenylalanine, valine, tyrosine, alanine, threonine, glutamic acid, serine, etc. The fruit is a good source of B vitamins and a fair source of ascorbic acid. Bitter fruits yield of solid foam containing cucurbitacins B, D, G and H, mainly cucurbitacin B; these bitter principles are present in the fruit as aglycones. The plant has various pharmacological activities like antioxidant, antihyperglycemic, antihyperlipidemic, cardio protective, immunomodulatory effects, hepato protective, in hyperthyroidism, hyperglycemia and lipid peroxidation, analgesic and anti-inflammatory, diuretic, cytotoxic activity.

CUCURBITACIN B

20. ***Lawsonia inermis* (Mehandi):** The leaves of the plant belonging to the family Lythraceae is reported to contain carbohydrates, proteins, flavonoids, tannins and phenolic compounds,

alkaloids, terpenoids, quinones, coumarins, xanthones and fatty acids. The plant has been reported to have analgesic, hypoglycemic, hepatoprotective, immunostimulant, anti-inflammatory, antibacterial, antimicrobial, antifungal, antiviral, antiparasitic, antitrypanosomal, antidermatophytic, antioxidant, antifertility, tuberculostatic and anticancer properties due to its active constituent lawsome .

LAWSONE

21. **Curculigo orchioides** (**Kalimusli**):   It consists of dried rhizome of *Curculigo orchioides* Gaertn family Amaryllidaceae. Its major constituents are Flavone, glycosides 5, curculigo saponins, hentria contanol, alkaloid lycorine, 2-methoxy-4-acetyl-5-methyltriacontane and behenic acid. It cures different sexual disorders in men such as low sperm count, piles, blood related disorders, skin disorders, jaundice, gonorrhoea, joint pain. The roots of this plant are a nice   stimulant,   appetizer,   carminative,   tonic   and   aphrodisiac.

LYCORINE

BEHENIC ACID

22. **Calotropis procera** (**Aak**): Arka consists of dried roots of *Calotropis procera* family Asclepiadaceae. Its major constituents are glycosides (calotropin). The leaves contain

ascorbic acid, calatropagenin and root has benzolisolineolone. The root skin, latex, flowers, leaves and the ksara of arka are used for medicinal purpose. The poultice of its leaves effectively reduces the pain and swelling in rheumatic joints and filariasis. The medicated oil is beneficial in otitis and deafness. The topical sprinkle of dried leaves powder hastens the wound healing. In glandular swellings the topical application of latex reduces the inflammation. In skin diseases, associated with depigmentation, the latex combined with mustard oil, works well. The fomentation with its leaves, slightly warmed with thin coat of castor oil, is beneficial to relieve the abdominal pain. The local application of latex is recommended in hairfall and baldness. It also, is useful in piles. The latex also mitigates the dental aches. The latex as a strong purgative and accumulations breaking imparts excellent results in ascites of kapha type and hepatosplenomegaly with ascites. To alleviate the edema in such conditions, of kapha origin, the decoction of its roots combined with triphala and honey, is salutary. In asthma and cough, the flowers and the root skin of arka are commonly used. As a blood purifier, it is benevolent is filariasis and syphilis. The red flowers alleviate raktapitta. In chronic dermatoses, the root skin is recommended with honey.

CALOTOPIN

23. *Tinospora cordifolia* (Giloe): It consists of dried, matured pieces of stem of Tinospora cordifolia family Menispermaceae. Its major constituents are terpenoids and alkaloids. The stem contains alkaloidal constituents, including berberine; bitter principles, including columbin, chasmanthin, palmarin and tinosporin, tinosporic acid and tinosporol. It is useful herbal as hepatoprotective and act as a remedy for infections, recurrent fevers also it acts as an immunomodulator useful in low immunity. It is useful for cancer of all types, high uric acid and in flu of all types.

BERBERINE

COLUMBIN

CHASMANTHIN

PALMARIN

24. ***Solanum surattense, Solanum xanthocarpum* (Kantkari):** It consists of mature, dried whole plant of *Solanum surattense, Solanum xanthocarpum* family Solanaceae. Its major constituents are glucoalkaloids and sterols. Fruits give solasonine, solamargine,

betasolamargine, and solasodine; petals yielded apigenin; stamens gave quercetin diglycoside and sitosterol. (+)- solanocarpine, carpesterol, solanocarpidine, potassium nitrate, fatty acid, diosgenin, sitosterol, isochlorogenic acid, neochronogenic acid, chronogenic acid, caffeic acid, solasodine, solasonine, solamargine, quercetin, apigenin, histamine, acetylcholine. Kantkari is useful in treating worms, cold, hoarseness of voice, fever, dysuria, enlargement of the liver, muscular pain, spleen and stone in the urinary bladder. Nasal administration of kantkari is beneficial in migraine, asthma and headache. The juice of the berries is used in curing sore throat. The fumigation of kantkari is helpful in piles. The herb is made to a paste and applied on swollen and painful joints to reduce the pain and swelling in arthritis. Roots and seeds are used as an expectorant in asthma, cough and pain in chest. The root is ground to a paste and mixed with lemon to cure snake and scorpion bites. Its stem, flowers and fruits, being bitter and carminative, are used for relieving burning sensation in the feet. Kantkari fruits also facilitate seminal ejaculation, alleviate worms, itching, and fever and reduce fats. The fruit works as an aphrodisiac in males. Its seeds are helpful for treating irregular menstruation and dysmenorrheal in females. The herb is beneficial in the treatment of cardiac diseases associated with edema, since it is a stimulant to the heart and a blood purifier.

SOLASONINE

APIGENIN

QUERECTIN

25. *Aloe barbadensis* (**Ghritkumari**): It consists of dried juice of leaves of *Aloe barbadensis* family Liliaceae. The major constituents are Anthraquinone glycoside- Aloe emodin, aloetic acid, anthrol, aloin A and B, isobarbaloin, emodin, ester of cinnamic acid. Dried juices of leaves are used in dysmenorrhoea and diseases of liver. It is used in jaundice due to viral hepatitis. It is also useful in spleen disorders. Gel topically is emollient, anti-inflammatory, antimicrobial used for wound healing, sunburn. *Aloe vera* detoxifies the body and is considered as best colon cleanser. It prevents constipation, controls diabetes, clear acne and skin allergies, dark spots.

EMODIN

ALOIN A

ALOIN B

ISOBARBALOIN

26. **Euphorbia neriifolia (Thuhar):** It consists of stem of *Euphorbia neriifolia* family Euphorbiaceae. The major constituents are resin, gum and triterpenes. The triterpenoids, euphol, 24- methylenecycloartenol, euphorbol hexacosonate, taraxerol, glut-5(10)-en-1-one, glut-5-en-3-beta-yet-acetate, friedelan-3-alphaol and -3-beta-ol have been reported. Latex used as purgative, diuretic, antiasthmatic, expectorant, rubefacient. It is used in ascites, polyuria, anasarca, chlorosis, tympanitis, externally warts, cutaneous eruptions, scabies, unhealthy ulcers. It is used as drastic purgative in the enlargement of liver and spleen, syphilis, dropsy, leprosy, etc.

TARAXEROL

27. *Phyllanthus niruri* (**Bhui amala**): It consists of root, stem and leaf of *Phyllanthus niruri* family Euphorbiaceae. The major constituents are lignans, alkaloids and bioflavonoids. The antihepatotoxic activity of Phyllanthus was attributed to two compounds in the plant called phyllanthin and hypophyllanthin. It is used in liver disorders and hepatitis B virus. It is also used as diuretic, deobstruent, astringent, anti-inflammatory, styptic. It is used in prescriptions for dyspepsia, indigestion, chronic dysentery, urinary tract diseases, diabetes, and skin eruptions.

PHYLLANTHIN

HYPOPHYLLANTHIN

28. *Calotropis gigantean* (**Madar**): It consists of dried root and bark of *Calotropis gigantean*, Family: Ascelpiadaceae. The root contains glycosides 0.60-1.42 % on dry basis. The latex contains akudarin. Flowers contain beta-amyrin and stigmasterol. The plant is purgative, alexipharmatic, antihelmintic, cureleprosy, leucoderma, ulcers, and piles, diseases of the spleen, the liver and the abdomen. The juice is antihelmintic and laxative, cures piles and "Kapha". The root bark is diaphoretic, cures asthma, syphilis. The flower is sweet, bitter, antihelmintic, analgesic, astringent, cures inflammation, tumours, rat bite. The milk is bitter,

heating, oleaginous, purgative, cures leucoderma, diseases of abdomen.

BETA-AMYRIN

STIGMASTEROL

29. ***Tephrosia purpurea* (Biyani):** It is a perennial herb obtained from *Tephrosia purpurea* belongs to a family Papilonaceae. The leaves contain rutin and rotenoids, triterpenoids, lupeol. Seeds contain a diketone-pongamol, a flavanone purpurin and sitosterol. The roots gave a prenylated flavanone 7-methylglabranin. Pods contain rotenoids such as villosin, villon, villosil, villosinol, villinol and villosone. Dried herb is diuretic, deobstruent, laxative. It is given for the treatment of cough, bronchitis, bilious febrile attacks, insufficiency of the liver, jaundice, kidney disorders and for the treatment of bleeding piles, boils, and pimples.

RUTIN

ROTENOIDS

.LUPEOL

SITOSTEROL

30. **Capparis deciduas (Karer):** It is a fruit obtained from *Capparis deciduas* family Capparidaceae. The root bark contains spermidine alkaloids. Stachydrine, glucobrassicin, glucocapparin and glucocleomin. The bark is bitter and diuretic. It is given in hepatic, spleen and renal complaints. It is used as anti-inflammatory used for enlarged cervical glands, sciatica, rheumatoid arthritis, externally on swelling. Fruits and seeds are used for urinary purulent discharges and dysentery and antimicrobial.

STACYDRINE

UCOBRASSICINGL

31. **Tecomella undulate (Rohiro):** It is fruit obtained from *Tecomella undulate* belongs to a family: Bignoniaceae. The bark conatins teconin, alkanes, alkanols, and beta-sterols. The bark also yielded chromone glycosides such as undulatosides A and B, and iridoid glycosides such as tecomelloside and tecoside. A quinonoid such as lapachol, vatic and dehydrotectol are also reported from the bark. Bark is used as relaxant, cardiotonic, choleretic. It is used for treatment for leucorrhoea, diseases of liver and spleen, leucoderma, syphilis and other skin diseases.

LAPACHOL

32. **Peganum harmala (Harmal):** It is perennial herb obtained from *Peganum harmala* family Nitrariaceae. Its major constituents are harmone, harmine, harmaline, harmalol, vasicine, vasicinone. The plant contains flavonoids such as kaempferol, quercetin and acacetin. Arial parts and seeds contain alkaloids like harmine, harmaline, harmalol. Inhalation of the smoke relieves pain in the liver. It is also employed in jaundice, asthma and colic. The powder of the seeds and watery infusion are given for the treatment of these diseases. The alkaloids exhibit antibacterial and antifungal activity.

HARMINE   ACACETIN

33. **Leucas aspera (Paniharin):** It is obtained from *Leucas aspera* family Lamincea. The plant gives oleanolic acid, ursolic acid and beta-sitosterol. The root contains a terpenoid, leucolactone and the sterols, sitosterols, stigmasterol and campesterol. It is used in jaundice, anorexia, dyspepsia, fever, helmintic manifestation, respiratory and skin diseases. Leaves are used as an external application for psoriasis, chronic skin eruption and painful swellings.

OLEANOLIC ACID

URSOLIC ACID

SITOSTEROL

STIGMASTEROL

34. *Cynara scolymus* (**Globe Artichoke**): The leaf of *Cynara scolymus* is used in Europe as a traditional medicine with choleretic, cholagogue and laxative properties to stimulate appetite and to treat liver insufficiency and hypercholesterolaemia. In France it is regarded as a liver tonic and hepatoprotective ('wringing out of the hepatic sponge') and as a depurative. The Quechua community of northern Bolivia uses an infusion of Globe Artichoke leaf for cirrhosis of the liver and colic caused by gallstones. Important constituents include the bitter tasting sesquiterpene, lactone, cynaropicrin as well as caffeic acid derivatives (including cynarin) and flavonoids.

CYNARPICRIN

35. **Taraxacum officinalis (Dandelion Root):** *Taraxacum officinalis* root is a choleretic, cholagogue, bitter tonic and mild laxative herb used traditionally for liver and gallbladder disorders such as inflammation of the gallbladder, gallstones, jaundice, dyspepsia with constipation and chronic skin conditions and its active constituent is taraxerol.

TARAXEROL

36. **Chionanthus virginica (Fringe Tree):** A phytochemical comparison of *Chionanthus virginica* has found that similar major compounds (especially lignans and secoiridoids) are present in the root bark and the stem bark. Although the ratios of these constituents vary, the stem bark provides a good substitute for the root bark. As the relative amount of major constituents is a little lower, an increase in the dosage of stem bark may be necessary. Fringe Tree root bark is a cholagogue, laxative and depurative herb. Popularised by the Eclectics, it has been used in western herbal medicine for poor appetite, dyspepsia, liver disease, jaundice, inflammation of the liver or gallbladder, bilious headache, enlargement of the liver or spleen and skin and bowel disorders due to reduced or disordered liver function.

37. **Allium cepa (Onion):** *Allium cepa* belongs to the family Liliaceae. Most onion cultivars are about 89% water, 4% sugar, 1% protein, 2% fibre and 0.1% fat. They contain vitamin

C, vitamin B_6, folic acid and numerous other nutrients in small amounts. They are low in fats and in sodium. Onion bulb contains chemical compounds such as phenolics and flavonoids that basic research shows to have potential anti-inflammatory, anti-cholesterol, hepato-protective, anticancer and antioxidant properties. These include quercetin and its glycosides quercetin-3, 4'-diglucoside and quercetin-4'-glucoside.

FOLIC ACID

VITAMIN B6

38. ***Ginkgo biloba* (Maidenhair tree):** *Ginkgo biloba* belongs to the family Ginkgoaceae. Extracts of ginkgo leaves contain flavonoids glycosides (myricetin and quercetin) and terpenoids (ginkgolides, bilobalides) and have been used pharmaceutically. These extracts are shown to exhibit reversible, nonselective monoamine oxidase inhibition, as well as inhibition of reuptake at the serotonin, dopamine, and norepinephrine transporters, with all but the norepinephrine reuptake inhibition fading in chronic exposure. Ginkgo extract has in addition been found to act as a selective 5-HT1A receptor agonist *in vivo*. *Ginkgo* supplements are usually taken in the range of 40–200 mg per day. In 2010, a meta-analysis of clinical trials has shown *Ginkgo* to be moderately effective in improving cognition in dementia patients but not preventing the onset of Alzheimer's disease in people without dementia. Ginkgo is believed to have no tropic properties, and is mainly used as memory and concentration enhancer, and antivertigo agent. Ginkgo extract may have three effects on the human body: improvement in blood flow to most tissues and organs, protection

against oxidative cell damage from free radicals, and blockage of many of the effects of platelet-activating factor (platelet aggregation, blood clotting) that have been related to the development of a number of cardiovascular, renal, respiratory and central nervous system disorders. Ginkgolides, especially ginkgolideB, are potent antagonists against platelet-activating factor, and thus may be useful in protection and prevention of thrombus, endotoxic shock, and from myocardial ischemia. The plant also contains biflavones important constituents present in the medicinally used leaves are the terpene trilactones, i.e., ginkgolides A, B, C, J and bilobalides, many flavonol glycosides, biflavones, proanthocyanidins, alkylphenols, simple phenolic acids, 6-hydroxykynurenic acid, 4-O-methylpyridoxine and polyprenols.

GINKGOLIDE A

39. *Momordica charantia* (Bitter melon): *Momordica charantia* belongs to the family Cucurbitaceae. Bitter melon fruit contains an array of biologically active plant chemicals including triterpenes, proteins, and steroids. One chemical has clinically demonstrated the ability to inhibit the enzyme guanylate cyclase that is thought to be linked to the cause of psoriasis and also necessary for the growth of leukemia and cancer cells. In addition, a protein found in bitter melon, momordin, has clinically demonstrated anticancerous activity against Hodgkin's lymphoma in animals. Other proteins in the plant, alpha- and beta-momorcharin and cucurbitacin B, have been tested for possible anticancerous effects. A chemical analog of these bitter melon proteins has been developed, patented, and named "MAP-30"; its developers reported that it was able to inhibit prostate tumor growth. Two of these proteins-alpha- and beta-momorcharin-have also been reported to inhibit HIV virus in test tube studies. In numerous studies, at least three different groups of constituents found in all parts of bitter melon have clinically demonstrated hypoglycemic (blood sugar lowering) properties or other actions of potential benefit against diabetes mellitus. These chemicals that

lower blood sugar include a mixture of steroidal saponins known as charantins, insulin-like peptides, and alkaloids. Alkaloids, charantin, charine, cryptoxanthin, cucurbitins, cucurbitacins, cucurbitanes, cycloartenols, diosgenin, elaeostearic acids, erythrodiol, galacturonic acids, gentisic acid, goyaglycosides, goyasaponins, guanylate cyclase inhibitors, gypsogenin, hydroxytryptamines, karounidiols, lanosterol, lauric acid, linoleic acid, linolenic acid, momorcharasides, momorcharins, momordenol, momordicilin, momordicins, momordicinin, momordicosides, momordin, multiflorenol, myristic acid, nerolidol, oleanolic acid, oleic acid, oxalic acid, pentadecans, peptides, petroselinic acid, polypeptides, proteins, ribosome-inactivating proteins, rosmarinic acid, rubixanthin, spinasterol, steroidal glycosides, stigmasta-diols, stigmasterol, taraxerol, trehalose, trypsin inhibitors, uracil, vacine, v-insulin, verbascoside, vicine, zeatin, zeatin riboside, zeaxanthin, and zeinoxanthin are all found in bitter melon.

MOMORDICIN

40. **Ocimum sanctum (Tulsi):** *Ocimum sanctum* belongs to the family Lamiaceae. Some of the main chemical constituents of tulsi leaves are: oleanolic acid, ursolic acid, rosmarinic acid, eugenol, carvacrol, linalool, β-caryophyllene (about8%), β-elemene (c.11.0%), and germacrene D (about 2%). A variety of *in vitro* studies and animal studies have indicated some potential pharmacological properties of *Ocimum tenuiflorum* or its extracts. Recent studies suggest tulsi may be a COX-2 inhibitor, like many modern painkillers, due to its high concentration of eugenol. The fixed oil has demonstrated antihyperlipidemic and cardioprotective effects in rats fed a high fat diet .Some laboratory experiments on extracts of *Ocimum tenuiflorum* have indicated they may have potential in future pharmaceutical applications in the field of cancer treatment, and mitigating the effects of radiation exposure. Isolated *O. sanctum* extracts have some antibacterial activity against *E. coli, S. aureus* and *P. aeruginosa.*

.OLEANOLIC ACID

.URSOLIC ACID

ROSMARINIC ACID

EUGENOL

41. **Ricinus communis (Castor oil plant):** *Ricinus communis* belongs to the family Euphorbiaceae. Castor seed is the source of castor oil, which has a wide variety of uses. The seeds contain between 40% and 60% oil that is rich in triglycerides, mainly ricinolein. The seed contains ricin, a toxin, which is also present in lower concentrations throughout the plant. Castor oil has many uses in medicine and other applications. An alcoholic extract of the leaf was shown, in lab rats, to protect the liver from damage from certain poisons. Methanolic extracts of the leaves of *Ricinus communis* were used in antimicrobial testing against eight pathogenic bacteria in rats and showed antimicrobial properties. The extract was not toxic. The pericarp of castor bean showed central nervous system effects in

mice at low doses. At high doses mice quickly died. A water extract of the root bark showed analgesic activity in rats [99]. Antihistamine and anti-inflammatory properties were found in ethanolic extract of *Ricinus communis* root bark.

42. ***Rubiatinctorum* (Common madder):** *Rubia tinctorum* belongs to the family Rubiaceae. The plant's roots contain several polyphenolic compounds like 1,3Dihydroxyanthraquinone (purpuroxanthin), 1,4Dihydroxyanthraquinone (quinizarin), 1, 2, 4-Trihydroxyanthraquinone (purpurin) and 1,2-dihydroxyanthraquinone (alizarin). This latter gives its red colour to a textile dye known as Rose madder. It was also used as a colourant, especially for paint, that is referred to as Madder Lake. In one study, madder was found to have antimicrobial properties *in vitro*. In one animal study, madder was found to have antidiarrheal activity in rats.

QUINIZARIN

ALIZARIN

PURPURIN

43. ***Zingiber officinale* (Ginger):** *Zingiber officinale* belongs to the family Zingiberaceae. The characteristic odor and flavor of ginger rhizome is caused by a mixture of zingerone, shogaols and gingerols, volatile oils that compose one to three percent of the weight of fresh ginger. In laboratory animals, the gingerols increase the motility of the gastrointestinal tract and have analgesic, sedative, antipyretic and antibacterial properties. [6]-gingerol (1-[4'-hydroxy-3'-methoxyphenyl]-5-hydroxy-3-decanone) is the major pungent principle of ginger. Ginger contains up to three percent of a fragrant essential oil whose main constituents are sesquiterpenoids, with (-)-zingiberene as the main component. Smaller

amounts of other sesquiterpenoids (β-sesquiphellandrene, bisabolene and farnesene) and a small monoterpenoid fraction (β-phelladrene, cineol, and citral) have also been identified. The pungent taste of ginger is due to nonvolatile phenylpropanoid-derived compounds, particularly gingerols and shogaols, which form from gingerols when ginger is dried or cooked. Zingerone is also produced from gingerols during this process; this compound is less pungent and has a spicy-sweet aroma. According to the American Cancer Society, ginger has been promoted as a cancer treatment "to keep tumors from developing", but "available scientific evidence does not support this". In limited studies, ginger was found to be more effective than placebo for treating nausea caused by seasickness, morning sickness and chemotherap. Zingerone may have activity against enterotoxigenic *Escherichia coli* in enterotoxin-induced diarrhea in mice.

ZINGERONE

GINGEROL

44. ***Ficus carica* (Common Fig):** *Ficus carica* belongs to the family Moraceae. Figs have a laxative effect and contain many antioxidants. The fruits are a good source of flavonoids and polyphenols including gallic acid, chlorogenic acid, syringic acid, (+)-catechin, (−)-epicatechin and rutin. In one study, a 40-gram portion of dried figs (two medium size figs) produced a significant increase in plasma antioxidant capacity and show hepatoprotective activity.

GALLIC ACID

CHLOOGENIC ACID

.SYRINGIC ACID

RUTIN

45. **Carica papaya (Papaya):** *Carica papaya* belongs to the family Caricaceae. Papaya fruit is a source of nutrients such as provitamin A, carotenoids, vitamin C, folate and dietary fiber. Papaya skin, pulp and seeds also contain a variety of phytochemicals, including lycopene and polyphenols. In preliminary research, danielone, a phytoalexin found in papaya fruit, showed antifungal activity against *Colletotrichum gloesporioides*, a pathogenic fungus of papaya. In some parts of the world, papaya leaves are made into tea as a treatment for malaria. Antimalarial and antiplasmodial activity has been noted in some preparations of the plant, but the mechanism is not understood and no treatment method based on these results has been scientifically proven. In belief that it can raise platelet levels in blood, papaya may be used as a medicine for dengue fever. Papaya is marketed in tablet form to remedy for digestive problems. Papain is also applied topically for the treatment of cuts, rashes, stings and burns.

LYCOPENE

DANIELONE

46. **Adhatoda vasica (Vasaka):** *Adhatoda vasica* belongs to the family Acanthaceae. The chief quinazoline alkaloid vasicine is reported in all parts of the plant, the highest being in inflorescence. (The modern drug Bromhexin is the synthetic form of vasicine) It is a bitter bronchodilator, respiratory stimulant, hypotensive, cardiac depressant, uterotonic and abortifacient. An aqueous solution of vasicinone hydrochloride, when studied in mice and dogs, was found to potentiate the bronchodilatory activity of aminophylline also that of isoprenaline. Vasicinone exhibited smooth muscle-relaxant properties of airways. Alkaloids present in the plant showed significant protection against allergen-induced bronchial obstruction in guinea pigs. The leaves are found to activate the digestive enzyme trypsin. An extract of the leaves showed significant antifungal activity against ringworm. The leaf-juice is stated to cure diarrhoea and dysentery.

VASICINE

VASICINONE

47. **Matricaria chamomilla (Chamomile):** Matricaria chamomilla belongs to the family Asteraceae. One of the active ingredients of its essential oil is the terpene bisabolol. Other active ingredients include farnesene, chamazulene, flavonoids (including apigenin, quercetin, patuletin and luteolin) and coumarins. Research with animals suggests antispasmodic, anxiolytic, anti-inflammatory and some antimutagenic and cholesterol-lowering effects for chamomile. Chamomile has sped healing time of wounds in animals. It also showed some benefit in an animal model of diabetes. It is used to treat diarrhoea and nausea. *In vitro* chamomile has demonstrated moderate antimicrobial and

antioxidant properties and significant antiplatelet activity, as well as preliminary results against cancer. Essential oil of chamomile was shown to be a potential antiviral agent against herpes simplex virus type 2 (HSV-2) *in vitro*.

FARNESENE

BISABOLOL

CHAMAZULENE

48. ***Kalanchoe pinnata* (Leaf of life):** Kalanchoe pinnata belongs to the family Crassulaceae. Kalanchoe pinnata has been found to contain bufadienolide cardiac glycosides. Bufadienolide compounds isolated from Kalanchoe pinnata include bryophillin A which shows strong anti-tumor promoting activity and bersaldegenin-3-acetate and bryophillin C which are less active. Bryophillin C also showed insecticidal properties. Several studies have documented that leaf of life is antibacterial, antimicrobial, hepatoprotective, antiviral and antifungal. The plant is also said to have effective antihistamine and anaphylactic properties that might explain its traditional use for asthma, insect bites and stings. In recent research in Hawaii, leaf of life demonstrated noticeable effects on cancer tissue and confirmed powerful antimicrobial activity. Leaf of life also exhibited pain relieving and anti-diabetic properties in a study on mice in Africa. The reported immuno-suppressant properties of leaf of life might therefore be useful in treating conditions such as rheumatoid arthritis and lupus.

BUFADIENOLIDE

49. **Solanum trilobatum (Alarka):** *Solanum trilobatum* belongs to the family Solanaceae. *Solanum trilobatum* is an extensively used Indian traditional medicine to cure various human ailments. It was distributed throughout the southern parts of India. *S. trilobatum* is reported to cure numerous diseases viz., tuberculosis, respiratory problems and bronchial asthma. *S. trilobatum* was reported to harbour hepatoprotective activity, antimicrobial activity, antioxidant activity, cytotoxic activity, haemolytic activity, protective effect, immunomodulatory activity and anti-inflammatory properties. Phytochemical screening showed the presence of carbohydrates, saponins, phytosterols and tannins in leaf, whereas, stem possess carbohydrates, saponins, phytosterols, tannins, flavonoids and cardiac glycosides as major phytochemical groups. Alkaloides such as soladunalinidine and tomatidine were isolated from the leaf and stem of *Solanum* species. *S. trilobatum* contains chemical compounds like Sobatum, β-solamarine, solasodine, solaine, glycoalkaloid and diosogenin.

SOLASODINE

50. **Trigonella foenum graecum (Fenugreek):** Is an annual herb that belongs to the family Leguminosae. The seeds of fenugreek are commonly used as a spice in food preparations

due to the strong flavour and aroma. The seeds are reported to have restorative and nutritive properties. Fenugreek seeds have antioxidant activity and have been shown to produce beneficial effects such as neutralization of free radicals and enhancement of antioxidant apparatus. The protective effect of a polyphenolic extract of fenugreek seeds against ethanol-induced toxicity was investigated in human Chang liver cells. Ethanolic treatment suppressed the growth of Change liver cells and induced cytotoxicity, oxygen radical formation and mitochondrial dysfunction.

51. *Garcinia mangostana* (**Mangos teen**): *Garcinia mangostana* belongs to the family Guttiferae. It is a tropical evergreen tree and is an emerging category of novel functional foods sometimes called "super fruits" presumed to have a combination of appealing subjective characteristics, such as taste, fragrance and visual qualities, nutrient richness, antioxidant strength and potential impact for lowering risk of human diseases . The pericarps of *G. mangostana* have been widely used as a traditional medicine for the treatment of diarrhea, skin infection and chronic wounds in South East Asia for many years. These are the nature's most abundant sources of xanthones, which are the natural chemical substances possessing numerous bioactive properties that help to maintain intestinal health, neutralize free radicals, help and support joints and cartilage functions and promotes immune systems. These are extracted from the rind of mangos teen containing 95% xanthones also isoflavones, tannin and flavonoids. Treatment of hepatocellular carcinomas (liver cancer) with chemotherapy has generally been disappointing and it is most desirable to have more effective new drugs. The investigators extracted and purified 6 xanthone compounds from the rinds (peel) of the fruit of Garcinia mangostana, mangos teen fruit. The investigators tested this extract on 14 different human liver cancer cell lines.

XANTHONES

52. *Jatropha curcas* (**Purging nut**): *Jatropa curcas* belongs to family Euphorbiaceae. It is an evergreen shrub, indigenous to America, but cultivated in most parts of India. This

evergreen plant is common in waste places throughout India, especially on the Coromandel Coast and in Travancore; in the southern parts it is cultivated chiefly for hedges in the Konkan, and also in Malay Peninsula. Leaves are regarded as antiphrastic, applied to scabies; rubefacient for paralysis, rheumatism; also applied to hard tumours. Leaves also show antileukemic activity. Compounds that have been isolated from *Jatropha curcas* leaves include the flavonoids apigenin and its glycosides vitexin and isovitexin, the sterols stigmasterol, α -D-sitosterol and its α – D-glucoside. Methanolic fraction of leaves of *Jatropha curcas (MFJC)* was evaluated against hepatocellular carcinoma induced by Aflatoxin.

APIGENIN

VITEXIN

STIGMASTEROL

53. *Cassia roxburghii* **(Ceylon Senna):** *Cassia roxburghii* belongs to the family Caesalpiniaceae. Seeds of *Cassia roxburghii* DC have been used in ethnomedicine for various liver disorders for its hepatoprotective activity. The methanolic extract of *Cassia roxburghii* reversed the toxicity produced by ethanol CCl4 combination in dose dependent manner in rats. The extract at the doses of 250 mg/kg and 500 mg/kg are

comparable to the effect produced by Liv-52®, a well established plants based hepatoprotective formulation against hepatotoxins.

54. *Solanum nigrum* (**Kakamachi**): *Solanum nigrum* belongs to the family Solanaceae. Ayurveda, the drug is known as kakamachi. Aromatic water extracted from the drug is widely prescribed by herbal vendors for liver disorders. Although clinical documentation is scare as far as hepatoprotective activity is concerned, but some traditional practitioners have reported favorable results with powdered extract of the plant.

55. *Coccinia grandis* (**Ivy gourd**): *Coccinia grandis* belongs to the family Cucurbitaceae. In traditional medicine, fruits have been used to treat leprosy, fever, asthma, bronchitis and jaundice.The fruit possesses mast cell stabilizing, anti-anaphylactic and antihistaminic potential. In Bangladesh, the roots are used to treat osteoarthritis and joint pain. A paste made of leaves is applied to the skin to treat scabies. There is some research to support that compounds in the plant inhibit the enzyme glucose-6-phosphatase. Glucose-6-phosphatase is one of the key liver enzymes involved in regulating sugar metabolism. Therefore, ivy gourd is sometimes recommended for diabetic patients. Although these claims have not been supported, there currently is a fair amount of research focused on the medicinal properties of this plant focusing on its use as an antioxidant, anti-hypoglycemic agent, immune system modulator, etc.

56. *Annona squamosa* (**Sugar apple**): *Annona squamosa* belongs to the family Annonaceae. The diterpenoid alkaloid atisine is the main component of the root. Other constituents of *Annona squamosa* include oxophoebine, reticuline, atidine, histisine, hetisine, hetidine, heterophyllisine, heterophylline, heterlophylline, isoatisine, dihydroatisine, hetisinone benzoyl heteratisine and citronella oil. In US patent 4689232, Bayer AG patented the extraction process and molecular identity of squamocin. This molecule is known as an annonaceous acetogenin. Bayer also patented its use as a biopesticide. Many others have found other acetogenins in extracts of the seeds, bark, and leaves. Studies have revealed that the extracts of *Annona squamosa* exert hepatoprotective effect and the plant extract could be an effective remedial for chemical induced hepatic damage.

RETICULINE

ACETOGENIN

57. **Lepidium sativum (Garden cress):** *Lepidium sativum* belongs to the family Brassicaceae. Garden cress seeds, since ancient times, have been used in local traditional medicine of India. Garden cress seeds are bitter, thermogenic, depurative, rubefacient, galactogogue, tonic, aphrodisiac, ophthalmic, antiscorbutic, antihistaminic and diuretic. They are useful in the treatment of asthma, coughs with expectoration, poultices for sprains, leprosy, skin disease, dysentery, diarrhoea, splenomegaly, dyspepsia, lumbago, leucorrhoea, scurvy and seminal weakness. Seeds have been shown to reduce the symptoms of asthma and improve lung function in asthmatics. The seeds have been reported as possessing a hypoglycemic property and the seed mucilage is used as a substitute for gum arabic and tragacanth.

58. **Aegle marmelos (Bael):** *Aegle marmelos* belongs to the family Rutaceae. The Tamil Siddhars call the plant *koovilam* and use the fragrant leaves for medicinal purposes, including dyspepsia and sinusitis. A confection called *ilakam* is made of the fruit and used to treat tuberculosis and loss of appetite. It is used in Ayurveda for many purposes, especially chronic constipation. Aegeline (N-[2-hydroxy-2(4-methoxyphenyl) ethyl]-3-phenyl-2-propenamide) is a known constituent of the bael leaf and consumed as a dietarysupplement for a variety of purposes.

AEGELINE

59. **Morinda citrifolia (Noni):** *Morinda citrifolia* belongs to the family Rubiaceae. *M. citrifolia* fruit contains a number of phytochemicals, including lignans, oligo- and polysaccharides, flavonoids, iridoids, fatty acids, scopoletin, catechin, beta-sitosterol, damnacanthal, and alkaloids. The green fruit, leaves, and root/rhizomes were traditionally used in Polynesian cultures to treat menstrual cramps, bowel irregularities, diabetes, liver diseases, and urinary tract infections.

DAMNACANTHAL

CATECHINS

SITOSTEROL

60. **Cichorium intybus (Kasni):** *Cichorium intybus* belongs to the family Asteraceae. *Cichorium intybus* is a popular Ayurvedic remedy for the treatment of liver diseases. It is a part of polyherbal formulations used in the treatment of liver diseases. In preclinical studies an alcoholic extract of the *Cichorium intybus* was found to be effective against chlorpromazine induced hepatic damage in adult albino rats. A bitter glucoside, Cichorin has been reported to be the active constituent of the herb.

61. **Coptidis rhizoma (Huanglian):** *Coptidis Rhizoma* belongs to the family Ranunculaceae. The extract is prepared from the rhizome of Coptis chinensis (huang lian) Berberine is an active compound in *Coptidis Rhizoma* with multiple pharmacological activities including

antimicrobial, antiviral, anti inflammatory, cholesterol-lowering, hepatoprotective and anticancer effects.

BERBERINE

Conclusion: It was concluded from the whole literature survey that modern society has inherited knowledge from many cultures that herbal medicines are effective candidates against hepatic disorder. Natural products have been used as medicine in Ayurveda from a long time. Some medicinal plants are hepatoprotective/hepatogenic agents against hepatotoxicity caused by hepatotoxicant. Many researchers have been done and more remain to be done on their safe and effective use.

Reference:

1. Pandey G. Medicinal plants against liver diseases. International Research journal of pharmacy. 2011; 2 (5):115–21.

2. Pandey GP. Pharmacological studies of Livol® with special reference to liver function. MVSc & AH. thesis, JNKVV, Jabalpur, MP, India. 1980.

3. Pandey GP. Hepatogenic effect of some indigenous drugs on experimental liver damage. PhD. Thesis, JNKVV, Jabalpur. 1990.

4. Kumar CH, Ramesh A, Suresh Kumar JN, Mohammed Ishaq B. A review on hepatoprotective activity of medicinal plants. Int J Pharmaceu Sci Res. 2011; 2(3):501–15.

5. Tredway scott. An ayurvedic herbal approach to a healthy liver. CNI 1998; 6(16).

6. Leung A. Encyclopedia of Common Natural Ingredients Used in Food, Drugs and Cosmetics. In: John Wiley & Sons, New York. 1980; p. 313–14.

7. Ammon HPT, Wahl MA. Pharmacology of Curcuma longa. Planta Medica. 1991; 57:1–7.

8. Tyler VE, Brady LR, Robbers JE. Pharmacognosy, 9th edition. In: Lea and Febiger, Philadelphia, PA. 198; p. 247–8.

9. Obermeier MT, White RE, Yang CS. Effects of bioflavonoids on hepatic P450 activities. Xenobiotica. 1995; 25:575–84.

10. Cao J, Xu Y, Chen J, et al. Chemopreventive effects of green and black tea on pulmonary and hepatic carcinogenesis. Fundam Appl Toxicol. 1996; 29:244–50.

11. Ody P. The Complete Medicinal Herbal. Dorling Kindersley Ltd, London. 1993; 44.

12. Tyler VE, Brady LR, Robbers JE. Pharmacognosy, 9th edition. In: Lea and Febiger, Philadelphia, PA. 1988; p. 68–9.

13. Merck Index, 10th edition. In; Merck & Co, Rahway, NJ 1983; 43–7.

14. Leung A. Encyclopedia of Common Natural Ingredients Used in Food drugs and Cosmetics. In: John Wiley and Sons, New York, NY. 1980; p. 220–3.

15. Crance JM, Biziagos E, Passagot J, et al. Inhibition of Hepatitis A replication in vitro by antiviral compounds. J Med Virol. 1990; 31:155–60.

16. Baba M, Shigeta S. Antiviral activity of glycyrrhizin against varicella-zoster virus in vitro. Antiviral Res. 1987; 7:99–107.

17. Ito M, Nakashima H, Baba M, et al. Inhibitory effect of glycyrrhizin on the in vitro infectivity and cytopathic activity of the human immunodeficiency virus [HIV (HTLV-III/LAV)]. Antiviral Res. 1987; 7:127–37.

18. Pompei R, Flore O, Marccialis MA, et al. Glycyrrhizic acid inhibits virus growth and inactivates virus particles. Nature. 1979; 281:689–90.

19. Pompei R, Pani A, Flore O, et al. Antiviral activity of glycyrrhizic acid. Experientia. 1980; 36:304.

20. Thakur CP, Thakur S, Singh PK, et al. The Ayurvedic medicines Haritaki, Amla and Bahira reduce cholesterol-induced atherosclerosis in rabbits. Intl J Cardiol. 1988; 21:167–75. 21. Gilani AH, Janbaz KH, Shah BH. Esculetin prevents liver damage induced by paracetamol and CCL4. Pharmacol Res. 1998; 37(1):31–5.

22. Kapil A. Antihepatoxic effects of major diterpenoid constituents of Andrographis paniculata. Biochem Pharmacol. 1993; 46(1):182–5.

23. Koul IB, Kapil A. Evaluation of the liver protective potential of piperine, and active principle of black and long peppers. Planta Med. 1993; 59: 413–17.

24. Kapoor LD. Handbook of Ayurvedic Medicinal Plants. 1st ed. Washington: CRC Press. 2005; 87–8.

25. Peter SR, Tinto WF, Mcleans S, Reynolds WF, Yu M. Cassane diterpenes from C. bonducella. Phytochemistry. 1998; 47(6):1153–55.

26. Archana P, Tandan SK, Chandra S, Lal J. Antipyretic and Analgesic activity of C. bonducella seed. Phytotherapy Research. 2005; 19(5): 376–81.

27. Sharma SR, Dwivedi SK, Swarup D. Hypoglycaemic, antihyperglycaemic and hypolipidcmic activities of C. bonducella seeds in rats. Journal of Ethnopharmacology. 1997; 58(1):39–44.

28. Saeed MA, Sabir AW. Antibacterial Activity of C. bonducella seed. Fitoterapia. 2001; 72(7):807–9.

29. Kapoor LD. CRC Handbook of Ayurvedic Medicinal Plants. Boca Raton (FL): CRC Press. 1990. 30. Bakhru HK. Herbs That Heal, Orient Paperbacks. New Delhi: 25.

31. Arjuna. The Ayurvedic Pharmacopoeia of India. 1(2):17.

32. Biswas K, Bhattacharya G, Haldar. Evaluation of Analgesic and AntiInflammatory Activities of Terminalia Arjuna Leaf, Journal of Phytology Phytopharmacology. 2011; 3(1):33–8.

33. Gupta S, Goyle S. Antioxidant and hypocholesterolaemic effects of Terminalia arjuna tree-bark powder: a randomised placebo-controlled trial. The Journal of the Association of Physicians of India. 2001; 49: 231–5.

34. Bharani G Bhargava. Salutary effect of Terminalia Arjuna in patients with severe refractory heart failure. International Journal of Cardiology. 1995; 49 (3):191–9.

35. Ragavan, Krishna k. Antidiabetic effect of T. arjuna bark extract in alloxan induced diabetic rats. Indian Journal of Clinical Biochemistry. 2006; 21(2): 123–8.

36. Sinha M, Sil. Aqueous extract of the bark of Terminalia arjuna plays a protective role against sodium-fluoride-induced hepatic and renal oxidative stress. Journal of Natural Medicines. 2007; 61(3):251–60.

37. Devi Narayan, Vani Devi. Gastroprotective effect of Terminalia arjuna bark on diclofenac sodium induced gastric ulcer. Chemico-Biological Interactions. 2007; 167(1):71–83.

38. Greenlee H, Abascal K, Yarnell E, Ladas E. Clinical Applications of Silybum marianum in Oncology. Integrative Cancer Therapies. 2007; 6(2):158–65.

39. Kroll DJ, Shaw HS, Oberlies NH. Milk Thistle Nomenclature: Why It Matters in Cancer Research and Pharmacokinetic Studies. Integrative Cancer Therapies. 2007; 6(2):110–19.

40. Hogan Fawn S, Krishnegowda Naveen K, Mikhailova Margarita, Kahlenberg Morton S. Flavonoid, Silibinin, Inhibits Proliferation and Promotes Cell-Cycle Arrest of Human Colon Cancer. Journal of Surgical Research. 2007; 14(1):58–65.

41. Rainone, Francine. Milk Thistle. American Family Physician. 2005; 72(7): 1285–88.

42. Huseini HF, Larijani B, Heshmat, R, Fakhrzadeh H, Radjabipour B, Toliat T, et al. The efficacy of Silybum marianum (L.) Gaertn. (silymarin) in the treatment of type II diabetes: A randomized, double-blind, placebocontrolled, clinical trial. Phytotherapy Research. 2006; 20(12):1036–39.

43. Gordon Adam, Hobbs Daryl A, Bowden D Scott, Bailey Michael J, Mitchell Joanne, Francis Andrew JP, Roberts Stuart K. Effects of Silybum marianum on serum hepatitis C virus RNA, alanine aminotransferase levels and well-being in patients with chronic hepatitis C. Journal of Gastroenterology and Hepatology. 2006; 21(1 Pt 2):275–80.

44. Pugh, Becky (5 December 2008). The Best Hangover Remedies Tested. The Telegraph. Retrieved 2 December 2012.

45. Picrorhiza kurroa. Monograph. Altern Med Rev 2001.

46. Stuppner H, Wagner H. New cucurbitacin glycosides from Picrorhiza kurrooa. Planta Med. 1989.

47. Trivedi NP, Rawal UM. Hepatoprotective and antioxidant property of Andrographis paniculata (Nees) in BHC induced liver damage in mice. Indian J Exp Biol. 2001; 39(1):41–6.

48. Shukla B, Visen PK, Patnaik GK and Dhawan BN. Choiretic effect of Andrographolide in rats and guinea pigs. Planta Med. 1992; 58(2):146–9.

49. Warrier PK, Nambiar VPK, Ramankutty C. Foeniculum vulgare. In: Indian Medical Plants. Chennai. Orient. 1978; p. 3.

50. Simándi BDA, Rónyani E, Yanxiang G, Veress T, Lemberkovics È, Then M, et al. Supercritical carbon dioxide extraction and fractionation of Fennel oil. J Agric Food Chem. 1999; 47(1):1635–40.

51. Hanefi Ö, Serdar U, Irfan B, Ismail U, Ender E, Abdurrahman Ö, Zübeyir HS. Hepatoprotective effect of Foeniculum vulgare essential oil: A carbontetrachloride induced liver fibrosis model in rats. J. Lab Anim Sci. 2004; 3(1):168–72.

52. Khosla P, Gupta DD, Nagpal RK. Effect of Trigonella foenum graecum (Fenugreek) on serum lipids in normal and diabetic rats. International Journal of Pharmacology.1995; 27(1):89–93.

53. Kokate CK, Purohit AP, Gokhale SB. Pharmacognosy. 37th ed. Nirali Prakashan. 2006. p. 232, 233, 248, 249, 251.

54. Sangameswaran B, Reddy TC, Jayakar B. Phytother Res. 2008; 22(1): 124–6.

55. Joshi P, Dhawan V. Current Science. 2005; 89 4.

56. The Ayurvedic pharmacopoeia of India, Government of India, Ministry of Health and Family Welfare, Department of Ayush. Part-I, Vol-I, 6, 27, 38, 39, 60, 63, 70, 79.

57. The Ayurvedic pharmacopoeia of India, Government of India, Ministry of Health and Family Welfare, Department of Ayush Part-I, Volume-II,10, 56, 57.

58. Kare CP. Indian Medicinal Plants- an illustrated dictionary. Springerverlag berlin/Heidelberg. 2007; 75–663.

59. Kirtikar KR, Basu BD. Indian medicinal plants. 2nd ed. International book distributors. 2005; Vol-III: p. 1607-09, 1841–42, 2019–20.

60. Kirtikar KR, Basu BD. Indian medicinal plants.2nd ed. International book distributorsr.2005; Vol-I: p. 197–9, 456-8, 724–5.

61. http://www.goherbalremedies.com/product/liver-cirrhosis.(htm accessed on 27.08.2010). 62. Khan TI, Dular AK. Biodiversity conversation in the Thar Desert; with emphasis on Endemic and Medicinal plants. The Environmentalist. 2003; 23:137–44.

63. Upadhyay B, Roy S, Kumar A. Traditional uses of medicinal plants among the rural communities of Churu district in the Thar desert, India. Journal of Ethnopharmacology. 2007; 113(3):387–99.

64. Rocchietta S. Minerva Med. 1959; 50:612.

65. Leclerc H. Précis de Phytothérapie, 5th Ed. Masson, Paris. 1983.

66. Fernandez EC et al. Fitoterapia. 2003; 74:407.

67. Mills S, Bone K. Principles and Practice of Phytotherapy. Modern Herbal Medicine. In: Churchill Livingstone, Edinburgh. 2000.

68. Felter HW, Lloyd JU. King's American Dispensatory. 18th Ed, 3rd revision. 1905; Vol 1: reprinted Eclectic Medical Publications, Portland; 1983.

69. British Herbal Medicine Association's Scientific Committee. British Herbal Pharmacopoeia. BHMA, Bournemouth; 1983.

70. British Herbal Medicine Association. British Herbal Compendium, Vol 1. BHMA, Bournemouth; 1992.

71. Penman KG et al. Aust J Med Herbalism. 2008; 20:107.

72. Ellingwood F, Lloyd JU. American Materia Medica, Therapeutics and Pharmacognosy. 11th Edn. Naturopathic Medical Series: Botanical Vol 2. Eclectic Medical Publications, Portland, 1983. 73. Slimestad R, Fossen T, Vågen I. M. Onions: a source of unique dietary flavonoids. Journal of Agricultural and Food Chemistry. 2007; 55 (25): 10067–80.

74. Williamson G, Plumb GW, Uda Y, Price KR, Rhodes Michael JC. Dietary quercetin glycosides: antioxidant activity and induction of the anticarcinogenic phase II marker enzyme quinone reductase in Hepalclc7 cells. Carcinogenesis. 1997; 17 (11):2385–87.

75. Olsson M.E, Gustavsson K.E, Vågen I.M. Quercetin and isorhamnetin in sweet and red cultivars of onion (Allium cepa L.) at harvest, after field curing, heat treatment, and storage. Journal of Agricultural Food Chemistry. 2010; 58 (4):2323–30.

76. Yasuo O, Paul AF, Norihiro K, Katsuhiko N, Myricetin and quercetin, the flavonoid constituents of Ginkgobiloba extract, greatly reduce oxidative metabolism in both resting and Ca2+-loaded brain neurons. Brain Research. 1994; 635(1, 2):125–9.

77. Fehske CJ, Leuner K, Müller WE. Ginkgo biloba extract (EGb761®) influences monoaminergic neurotransmission via inhibition of NE uptake, but not MAO activity after chronic treatment. Pharmacological Research. 2009; 60 (1):68–73.

78. Winter JC, Timineri D. The discriminative stimulus properties of EGb 761, an extract of Ginkgo biloba. Pharmacology, Biochemistry, and Behavior. 1999; 62 (3):543–7.

79. Weinmann S, Roll S, Schwarzbach C, Vauth C, Willich SN. Effects of Ginkgo biloba in dementia: systematic review and meta-analysis. BMC geriatrics. 2010; 10:14.

80. Dekosky S. T, Williamson J. D, Fitzpatrick A. L, Kronmal R. A, Ives D. G, Saxton J. A, Lopez O. L, Burke G. et al. Ginkgo biloba for Prevention of Dementia: A Randomized Controlled Trial. JAMA: the Journal of the American Medical Association. 2008; 300(19):2253–62.

81. Snitz B. E, O›Meara E. S, Carlson M. C, Arnold A. M, Ives D. G, Rapp S. R, Saxton J, Lopez O. L. et al. Ginkgo biloba for Preventing Cognitive Decline in Older Adults: A Randomized Trial. JAMA: the Journal of the American Medical Association. 2009; 302(24):2663–70.

82. Mahadevan S, Park Y. Multifaceted Therapeutic Benefits of Ginkgo biloba L.: Chemistry, Efficacy, Safety, and Uses. Journal of Food Science. 2007; 73(1):R14–9.

83. Smith P, MacLennan K, Darlington CL. The neuroprotective properties of the Ginkgo biloba leaf: a review of the possible relationship to plateletactivating factor (PAF). Journal of Ethnopharmacology. 1997; 50 (3):131–9.

84. Ding-q, Chen J. Pharmacological Activities of Ginkgolides(School of Biological and Environmental Engineering, Jiangsu University of Science and Technology, Zhenjiang, Jiangsu 212013, China).

85. Pietta P, Mauri P, Rava A. Reversed-phase high-performance liquid chromatographic method for the analysis of biflavones in Ginkgo biloba L. extracts. Journal of chromatography. 1988; 437(2):453–6.

86. Teris A, van B. Chemical analysis of Ginkgo biloba leaves and extracts. Journal of Chromatography A. 2002; 967(1):21–55.

87. Zhu F et al. Alpha-momorcharin, a RIP produced by bitter melon, enhances defense response in tobacco plants against diverse plant viruses and shows antifungal activity in vitro. Planta. 2012.

88. Santos K, et al. Trypanocide, cytotoxic, and antifungal activities of Momordica charantia. Pharm Biol. 2012; 50(2):162–66.

89. Kuhn M, David W, Winston & Kuhn›s. Herbal Therapy & Supplements: A Scientific and Traditional Approach In: Lippincott Williams & Wilkins. p. 260.

90. Padalia RC, Verma RS. Comparative volatile oil composition of four Ocimum species from northern India. Natural Product Research. 2007; 25(6):569–75.

91. Prakash P, Gupta N. Therapeutic uses of Ocimum sanctum Linn (Tulasi) with a note on eugenol and its pharmacological actions: A short review. Indian Journal of Physiology and Pharmacology. 2005; 49(2):125–31.

92. Suanarunsawat T, Boonnak T, Na Ayutthaya WD, Thirawarapan S. Antihyperlipidemic and cardioprotective effects of Ocimum sanctum L. fixed oil in rats fed a high fat diet. Journal of Basic and Clinical Physiology and Pharmacology. 2011; 21(4):387–400.

93. Baliga MS, Jimmy R, Thilakchand KR, Sunitha V, Bhat NR, Saldanha E, Rao S, Rao P, Arora R, Palatty PL. Ocimum sanctum L (Holy Basil or Tulsi) and its phytochemicals in the prevention and treatment of cancer. Nutrition and cancer. 2013; 65 Suppl 1:26–35.

94. Golshahi H, Ghasemi E, Mehranzade E editors. Antibacterial activity of Ocimum sanctum extract against E. coli, S. aureus and P. aeruginosa. Clinical Biochemistry. Conference: 12th Iranian Congress of Biochemistry, ICB and 4th International Congress of Biochemistry and Molecular Biology. 2011; 44 (13 SUPPL. 1): S352.

95. Joshi M, Waghmare S, Chougule P, Kanase A. Extract of Ricinus communis leaves mediated alterations in liver and kidney functions against single dose of CCl4 induced liver necrosis in albino rats. Journal of Ecophysiology and Occupational Health. 2004; 4(3–4):169–73.

www.ingramcontent.com/pod-product-compliance
Lightning Source LLC
Chambersburg PA
CBHW081302180526

45170CB00007B/2534